PERENNIAL MATH

CONTEST PROBLEMS

2017-2018

Dr. Sylvia Dean was a gifted specialist and administrator for 26 years in Huntsville, Alabama. She created Perennial Math in 2011 upon completion of her doctoral degree. She has hosted many robotics, technology, and mathematics competitions throughout the years and has tirelessly worked on providing educational opportunities to students around the world. It has always been her ambition to leave students with the feeling that "math is cool."

These problems are challenging and students should know up front that scoring a 3 out of 6 on any of these tests is extraordinary and above average. Great problem-solvers learn persistence and appreciation for challenge. Good luck as you encourage your students to take the quest to become great mathletes.

Intermediate
level

Grades 5-6

Name________________________________

PERENNIAL MATH TEST #1

January 2017 Intermediate Level

1) How many numbers between 1-100 are equal to four times an odd number?	
2) What is the digit in the 10th position in the decimal representation of 1/17?	
3) Bill and Mark have 39 marbles together. Mark and Calvin have 41 marbles together. Bill and Calvin have 44 marbles together. How many marbles does Calvin have?	
4) This shape is made up of small cubes. Jim needed to make seven of these shapes for a model he was building. How many cubes did he need to build all seven of them? 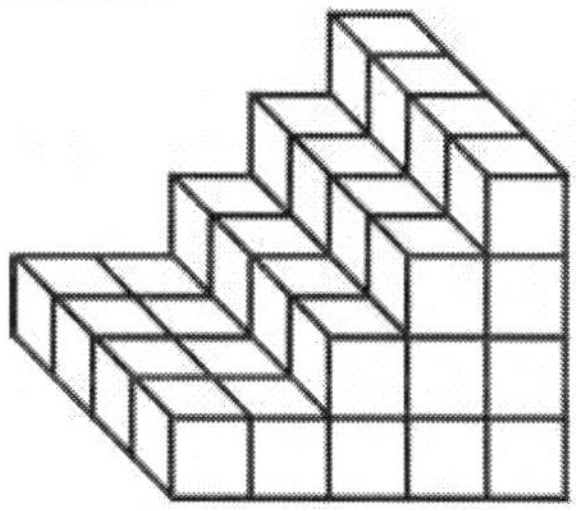	
5) When Mitzi counts her beads by fives she has 3 left over. When she counts them by sevens, she has 4 left over. What is the least prime number of beads she can have?	
6) If 10 men can cut 30 logs in 8 hours, how many logs will 40 men cut in 4 hours if they are working at the same rate?	

PERENNIAL MATH TEST #1

January 2017 Intermediate Level Answers

1) How many numbers between 1-100 are equal to four times an odd number?	12
2) What is the digit in the 10th position in the decimal representation of 1/17?	4
3) Bill and Mark have 39 marbles together. Mark and Calvin have 41 marbles together. Bill and Calvin have 44 marbles together. How many marbles does Calvin have?	23
4) This shape is made up of small cubes. Jim needed to make seven of these shapes for a model he was building. How many cubes did he need to build all seven of them? 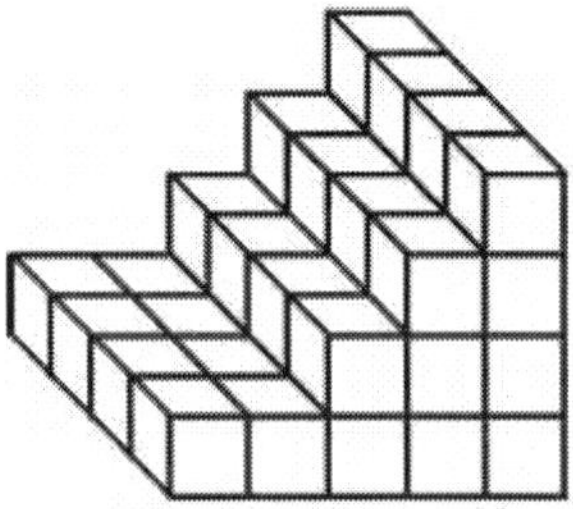	308
5) When Mitzi counts her beads by fives she has 3 left over. When she counts them by sevens, she has 4 left over. What is the least prime number of beads she can have?	53
6) If 10 men can cut 30 logs in 8 hours, how many logs will 40 men cut in 4 hours if they are working at the same rate?	60

PERENNIAL MATH

TEST # 1

January 2017 Intermediate Level Solutions

1) The numbers between 1-100 are equal to four times an odd number are: 4 x (1, 3, 5, 7, 9, 11, 13, 15, 17, 19, 21 and 23). Do not count 25 because the problem states numbers between 1 and 100 so 4 x 25 = 100 and it cannot be counted. There are 11 numbers.

12

2) 1 ÷ 17 = .0588235294117647. The digit in the 10th position in the decimal representation of 1/17 is 4.

4

3) Use algebra: Bill (B) and Mark (M) have 39 marbles together, B + M = 39. Mark (M) and Calvin (C) have 41 marbles together, M + C = 41. Bill and Calvin have 44 marbles together, B + C = 44. Solve for Calvin (C).

$$\begin{aligned} B + M \quad &= 39 \\ M + C &= 41 \\ \underline{B + \quad C} &\underline{= 44} \\ 2B + 2M + 2C &= 124 \quad \text{Divide both sides by 2} \\ B + M + C &= 62 \\ \underline{-(B + M \quad} &\underline{= 39)} \quad \text{Now subtract the known } (B + M = 39) \text{ out to solve for C} \\ C &= 23 \end{aligned}$$

23

PERENNIAL MATH

TEST # 1

January 2017 Intermediate Level Solutions

4) Find out how many cubes make up 1 shape. This bottom layer is 5 x 4 = 20 cubes. The next layer is 3 x 4 = 12 cubes. The next layer is 2 x 4 = 8 cubes. And the top layer is 4 cubes. That makes 44 cubes in the shape. Jim needed to make 7 of these so he needed 7 x 44 or 308 cubes.

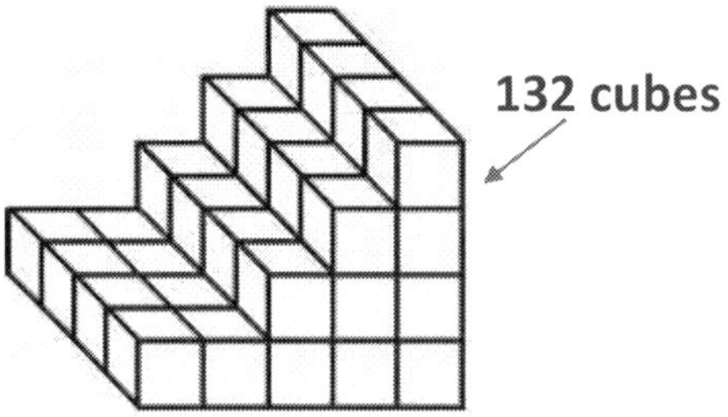

308

5) Make a chart to find the number of beads she could have and then which number is the least prime. When Mitzi counts her beads by fives she has 3 left over. When she counts them by sevens, she has 4 left over. She could have 18, but that is not prime. The least prime number of beads she can have is 53.

Counts by 5 with 3 left over	8	13	18	23	28	33	38	43	48	53
Counts by 7 with 4 left over	11	18	25	32	39	46	53			

53

6) This is a rate of work problem. Let t = number of trees that 40 men can cut in 4 hours. We know it takes 10 men 8 hours, or 10*8=80 man-hours to cut 30 trees. So, 40 men working only 4 hours would be 40*4=160 man-hours. Use a ratio to solve: 80 man-hours is to 30 trees as 160 man-hours is to t trees, or:

80/30 = 160/t cross multiply

80t = 30 * 160

80t = 4800

t = 60 trees

60

Name________________________________

PERENNIAL MATH

TEST #2

February 2017 Intermediate Level

1) 18 + ☐ = 26 - ☐

2) Farmer Brown only grows cabbages and cucumbers. He planted 6 cabbages for every 5 cucumbers. If Farmer Brown had 66 total vegetables that he put in his wagon, how many of those were cabbages?

3) It rained a lot in April so Bob could not play outside for 18 days. What percent of the month did it not rain?

4) In this magic square the four numbers in each row, column, and diagonal have the same sum. What value should X have?

1	15		
	6	7	X
8	10	11	
		2	16

5) Out of the 48 boys in the 5th grade twenty-one boys play baseball and twenty-five boys play football. Three boys play both sports. How many boys do not play either sport?

6) Mr. Johnson decided to invest his money this year. In January he invested $3 in a stock, then $9 in February, then $27 in March and he kept putting in triple the amount each month. How much money did he invest for the year?

PERENNIAL MATH

TEST #2

February 2017 Intermediate Level Answers

<table>
<tr><td>1) 18 + $\square$ = 26 - $\square$</td><td>4</td></tr>
<tr><td>2) Farmer Brown only grows cabbages and cucumbers. He planted 6 cabbages for every 5 cucumbers. If Farmer Brown had 66 total vegetables that he put in his wagon, how many of those were cabbages?</td><td>36</td></tr>
<tr><td>3) It rained a lot in April so Bob could not play outside for 18 days. What percent of the month did it not rain?</td><td>40</td></tr>
<tr><td>4) In this magic square the four numbers in each row, column, and diagonal have the same sum. What value should X have?
<table><tr><td>1</td><td>15</td><td></td><td></td></tr><tr><td></td><td>6</td><td>7</td><td>X</td></tr><tr><td>8</td><td>10</td><td>11</td><td></td></tr><tr><td></td><td></td><td>2</td><td>16</td></tr></table></td><td>9</td></tr>
<tr><td>5) Out of the 48 boys in the 5th grade twenty-one boys play baseball and twenty-five boys play football. Three boys play both sports. How many boys do not play either sport?</td><td>5</td></tr>
<tr><td>6) Mr. Johnson decided to invest his money this year. In January he invested $3 in a stock, then $9 in February, then $27 in March and he kept putting in triple the amount each month. How much money did he invest for the year?</td><td>797,160</td></tr>
</table>

PERENNIAL MATH

TEST # 2

February 2017 Intermediate Level Solutions

1) 18 + ☐ = 26 - ☐

The number that goes in the box that makes both sides equal is 4.
18 + 4 = 26 – 4. Both sides would equal 22.

4

2) This is a ratio problem. Since the ratio is 6:5 cabbages to cucumbers, then for 6 out of every 11 vegetables would have to be cabbages. Since there were 66 vegetables, then 66/11 = 6. Six portions would be 6 x 6 or 36 cabbages.

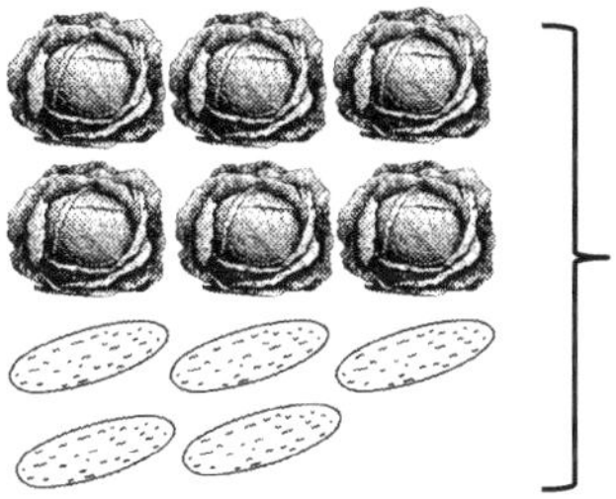

(6 of 11 are cabbages)
36 out of 66 are cabbages

36

3) It rained a lot in April so Bob could not play outside for 18 days. April has 30 days, so he could play outside for 12 days. Convert 12/30 to a percent to find out what percent of the month it did not rain. 12/30 = 2/5 which is 40%.

40

PERENNIAL MATH

TEST #2

February 2017 Intermediate Level Solutions

4) The sums of each row, column and diagonal are equal. You can go in any direction to solve for the sides and then finally you will have three numbers in the fourth column and that will allow you to solve for X. One of the diagonals has all 4 numbers so we can solve for the sum: 1 + 6 + 11 + 16 = 34. Now solve for the other empty squares. 15 + 6 + 10 + ? = 34. The missing number is 3. The lower left corner square is 3 + 2 + 16 + ? = 34. The number is 13. The top right corner is 13 + 10 + 7 + ? = 34. The number is 4. Solve for the third row: 8 + 10 + 11 + ? = 34. The missing number is 5. Now solve for X. 16 + 5 + 4 + X = 34. X = 9

1	15		4
	6	7	X
8	10	11	5
13	3	2	16

9

5) Create a Venn Diagram. Out of the 48 boys in the 5th grade twenty-one boys play baseball and twenty-five boys play football. Three boys play both sports. Subtract to find out how many boys fit in each region. There are 5 boys who did not play neither sport.

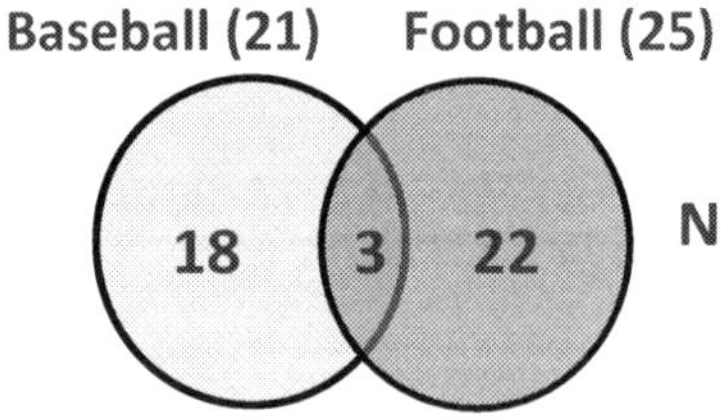

Neither = 48 - (18+3+22)
= 48 – 43
= 5

5

6) Make a chart to keep up with the money invested each month. Then add the amounts together.

January	3
February	9
March	27
April	81
May	243
June	729
July	2187
August	6561
September	19683
October	59049
November	177147
December	531441
Total	797160

797,160

Name________________________________

PERENNIAL MATH

TEST #3

March 2017 Intermediate Level

1) Evaluate 17 – 16 + 15 – 14 + · · · + 3 – 2 + 1.	
2) If ◊ + ◊ = 12; ⌂ + Δ = ◊; and ⌂+ ◊ = 11. Find the number that Δ represents.	
3) 4 Fuzzies are equal to 1 Wuzzie and 2 Wuzzies are equal to 1 Yuzzie. How many Fuzzies make up 8 Wuzzies and 4 Yuzzies?	
4) How many rectangles are in this shape? 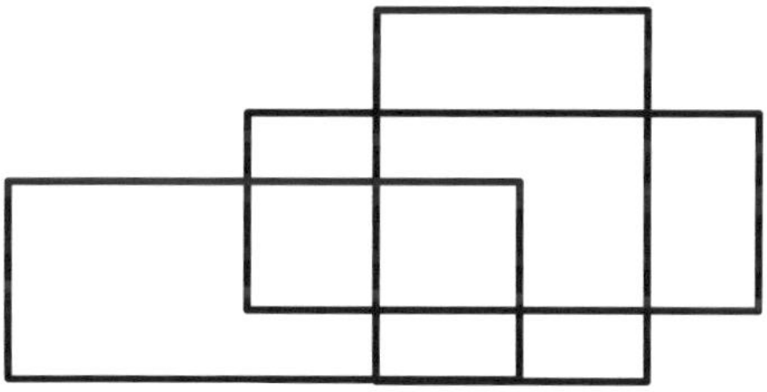	
5) The store is selling 12 oranges in a bag for $3.00. They also have bags containing 24 oranges and want to sell these so that each orange is 5 cents less than those in the bag containing 12 oranges. What should they charge for the bag that contains 24 oranges?	
6) The number 18 is equal to two times the sum of its digits 1 + 8 = 9, 2 x 9 = 18. Find the two-digit numbers that are equal to four times the sum of their digits. What is the sum of those numbers?	

PERENNIAL MATH TEST # 3

March 2017 Intermediate Level Answers

1) Evaluate 17 – 16 + 15 – 14 + · · · + 3 – 2 + 1.	9
2) If ◊ + ◊ = 12; ⌂ + Δ = ◊; and ⌂+ ◊ = 11. Find the number that Δ represents.	1
3) 4 Fuzzies are equal to 1 Wuzzie and 2 Wuzzies are equal to 1 Yuzzie. How many Fuzzies make up 8 Wuzzies and 4 Yuzzies?	64
4) How many rectangles are in this shape? 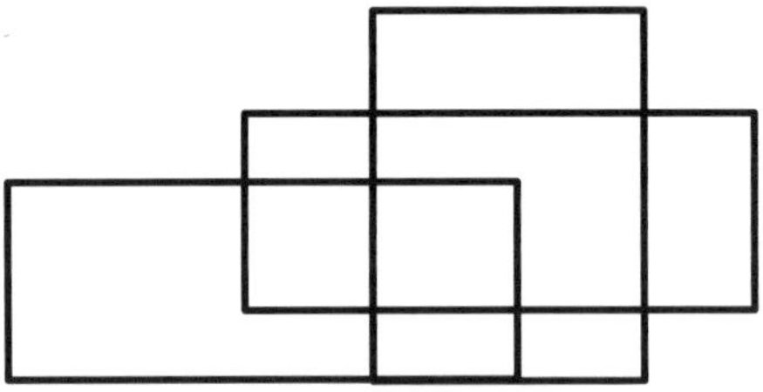	20
5) The store is selling 12 oranges in a bag for $3.00. They also have bags containing 24 oranges and want to sell these so that each orange is 5 cents less than those in the bag containing 12 oranges. What should they charge for the bag that contains 24 oranges?	4.80
6) The number 18 is equal to two times the sum of its digits 1 + 8 = 9, 2 x 9 = 18. Find the two-digit numbers that are equal to four times the sum of their digits. What is the sum of those numbers?	120

PERENNIAL MATH

TEST # 3

March 2017 Intermediate Level Solutions

1) Evaluate 17 – 16 + 15 – 14 + · · · + 3 – 2 + 1.

(17-16) + (15-14) + (13-12) + (11-10) + (9-8) + (7 – 6) + (5-4) + (3-2) + 1 =

1 + 1 + 1 + 1 + 1 + 1 + 1 + 1 + 1 = 9

9

2) If ◊ + ◊ = 12; ⌂ + Δ = ◊; and ⌂+ ◊ = 11. Find the number that Δ represents.

Solve for ◊: ◊ + ◊ = 12

◊ = 6

Solve for ⌂: ⌂ + ◊ = 11

⌂ + 6 = 11

⌂ = 5

Substitute ◊ and ⌂ into: ⌂ + Δ = ◊

5 + Δ = 6

Δ = 1

1

3) 4 Fuzzies are equal to 1 Wuzzie and 2 Wuzzies are equal to 1 Yuzzie. To find out how many Fuzzies make up 8 Wuzzies and 4 Yuzzies first solve the original equations and then plug in the values.

4F = W

2W = Y Therefore, 8F = Y

8W = (8x4)F

8W = 32F

4Y= (4 x 8)F

4Y = 32F

So there are 32F + 32F = 64 F in 8W and 4Y.

64

PERENNIAL MATH

TEST #3

March 2017 Intermediate Level Solutions

4) Try to count the rectangles in a consistent manner by counting individual shapes, then rectangles made from 2 shapes, 3 shapes, etc. Then use a method of counting across and then down. There are 20 different rectangles in this shape.

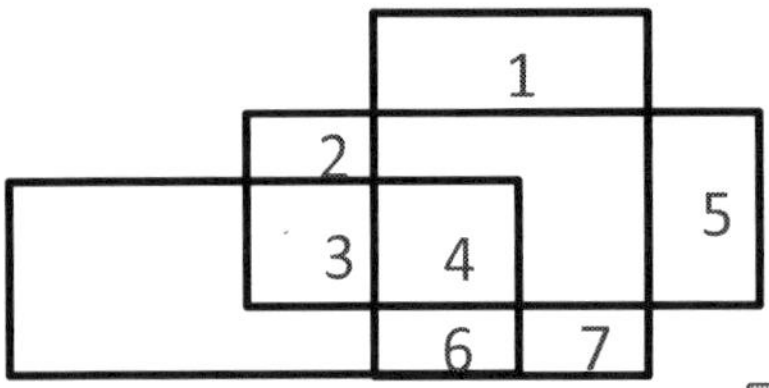

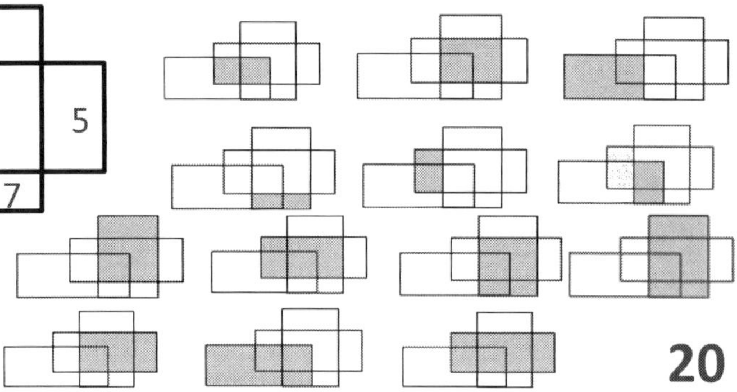

20

5) The store is selling 12 oranges in a bag for \$3.00. They also have bags containing 24 oranges and want to sell these so that each orange is 5 cents less than those in the bag containing 12 oranges. To find what they should charge for the bag that contains 24 oranges first solve for what they are charging for each orange in the bag of 12. \$3 ÷ 12 = .25 each. So if they want to sell the oranges for .05 less, then a bag of 24 will cost 24 x .20 = \$4.80

4.80

6) First find the two-digit numbers that are equal to four times the sum of their digits. Those numbers are 12, 24, 36, 48. 1+2 = 3, 2 + 4 = 6, 3 + 6 = 9, and 4 + 8 = 12 and 4 x 3 = 12, 4 x 6 = 24, 4 x 9 = 36 and 4 x 12 = 48 respectively. The sum of these number is 12 + 24 + 36 + 48 = 120.

120

Name________________________________

PERENNIAL MATH

TEST #4

April 2017 Intermediate Level

1) $2 \cdot 5 \cdot 6 + 3 - 2 \div 2$	
2) Smitty bought a bag of popcorn and a box of candy for $2.00. He could have purchased 2 bags of popcorn and no candy for $1.50. How much would he pay for 5 bags of popcorn and 10 boxes of candy?	
3) Andrew has three nickels, three dimes, and two quarters. How many different amounts of money can he make using one or more of the coins?	
4) In this cryptarythm, each letter stands for a different digit. What is the largest value that ACA can be? $$\begin{array}{r} AA \\ \times\ BB \\ \hline ACA \end{array}$$	
5) Farmer Brown and Farmer Gray have 88 chickens together. If 6 chickens are transferred over from Farmer Gray to Farmer Brown, then Farmer Brown will have as many chickens as Farmer Gray. How many chickens does Farmer Gray have before the transfer?	
6) Jessica has more than 100 beads. If she divides her beads equally among 6, 7, or 8 friends she will have one bead left over. What is the least number of beads that Jessica could have?	

PERENNIAL MATH

TEST #4

April 2017 Intermediate Level Answers

1) $2 \cdot 5 \cdot 6 + 3 - 2 \div 2$	62
2) Smitty bought a bag of popcorn and a box of candy for $2.00. He could have purchased 2 bags of popcorn and no candy for $1.50. How much would he pay for 5 bags of popcorn and 10 boxes of candy?	16.25
3) Andrew has three nickels, three dimes, and two quarters. How many different amounts of money can he make using one or more of the coins?	19
4) In this cryptarythm, each letter stands for a different digit. What is the largest value that ACA can be? $$\begin{array}{r} AA \\ \underline{\times\ BB} \\ ACA \end{array}$$	484
5) Farmer Brown and Farmer Gray have 88 chickens together. If 6 chickens are transferred over from Farmer Gray to Farmer Brown, then Farmer Brown will have as many chickens as Farmer Gray. How many chickens does Farmer Gray have before the transfer?	50
6) Jessica has more than 100 beads. If she divides her beads equally among 6, 7, or 8 friends she will have one bead left over. What is the least number of beads that Jessica could have?	169

PERENNIAL MATH TEST #4

April 2017 Intermediate Level Solutions

1) Use the order of operations to solve:

$$2 \cdot 5 \cdot 6 + 3 - 2 \div 2 = (2 \times 5 \times 6) + 3 - (2 \div 2) =$$
$$60 \quad + 3 - \quad 1 \quad = 62$$

62

2) To solve for how much he would pay for 5 bags of popcorn and 10 boxes of candy, first solve for the costs for 1 bag of popcorn and 1 box of candy. Since he could have purchased 2 bags of popcorn and no candy for \$1.50, a bag of popcorn costs 1.50 ÷ 2 = .75. Subtract this from 2.00 since Smitty bought a bag of popcorn and a box of candy for \$2.00, and a box of candy costs 1.25. Now multiply to find the answer.

5 (\$0.75) + 10 (\$1.25) = \$3.75 + \$12.50 = \$16.25

16.25

3) Andrew has three nickels, three dimes, and two quarters. How many different amounts of money can he make using one or more of the coins? Make a chart or ordered list. There are 19 different amounts he could make.

Quarters	Dimes	Nickels	Amount
2	3	3	0.95
2	3	2	0.9
2	3	1	0.85
2	3	0	0.8
2	2	3	0.85
2	2	2	0.8
2	2	1	0.75
2	2		0.7
2	1	3	0.75
2	1	2	0.7
2	1	1	0.65
2	1	0	0.6
2	0	3	0.65
2	0	2	0.6
2	0	1	0.55
2	0	0	0.5
1	3	3	0.7
1	3	2	0.65
1	3	1	0.6
1	3	0	0.55
1	2	3	0.6
1	2	2	0.55
1	2	1	0.5
1	2		0.45
1	1	3	0.5
1	1	2	0.45
1	1	1	0.4
1	1	0	0.35
1	0	3	0.4
1	0	2	0.35
1	0	1	0.3
1	0	0	0.25
0	3	3	0.45
0	3	2	0.4
0	3	1	0.35
0	3	0	0.3
0	2	3	0.35
0	2	2	0.3
0	2	1	0.25
0	2		0.2
0	1	3	0.25
0	1	2	0.2
0	1	1	0.15
0	1	0	0.1
0	0	3	0.15
0	0	2	0.1
0	0	1	0.05

19

PERENNIAL MATH

TEST #4

April 2017 Intermediate Level Solutions

4) A and B can only be 1 or 2 because the product is a three-digit number. Now you can do the trial and error method to see if the product is equal to ACA. The number ACA can only be 242, 363 or 484. The largest number that ACA can be is 484.

AA	22	33	44
x BB	x 11	x 11	x 11
ACA	242	363	484

484

5) Farmer Brown and Farmer Gray have 88 chickens together. If 6 chickens are transferred over from Farmer Gray to Farmer Brown, then Farmer Brown will have as many chickens as Farmer Gray. That means they will both have 88/2 = 44 chickens. So Farmer Gray had 44 + 6 = 50 chickens to start with and Farmer Brown had 44 – 6 = 38 chickens.

50

6) Jessica has more than 100 beads. If she divides her beads equally among 6, 7, or 8 friends she will have one bead left over. To find the least number of beads that Jessica could have find the LCM of 6, 7 and 8 and then add 1 to it. The LCM of 6, 7 and 8 is 168, and 168 +1 left over is 169.

169

Name________________________________

PERENNIAL MATH

TEST #1

November 2017 Intermediate Level

1) The third number is made by adding the previous 2 numbers. What number will go in the last box?

0.8		2.3		

2) The science test has some 3-point and some 4-point questions and 100 is the highest score you can make. If there are four more 4 point questions than 3-point questions, how many 3-point questions are on the test?

3) Mike and Bill are cousins whose ages are the same two different digits, only reversed. The sum of their ages is 132. Mike is less than 20 years older than Bill. How old is Bill?

4) Use the numbers 1-6 in the circles so that the sum along each side of the triangle equals 11. What is the sum of the three corner circles?

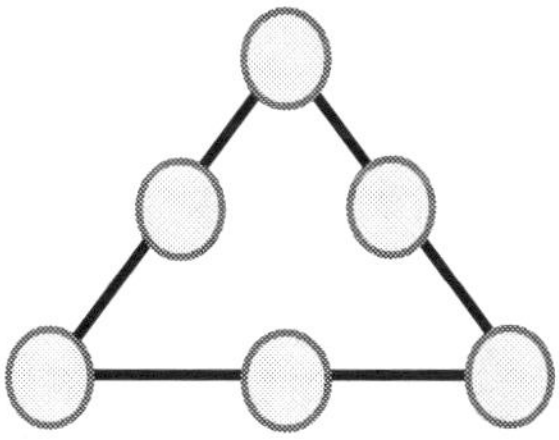

5) How many ways can 20 be written as the sum of 3 different numbers?

6) Tracy has 50 pennies and 6 cups. She plans to put an odd number of pennies in each cup. No cup can contain a number that is divisible by 5 pennies because she doesn't like that number. What is the most amount of pennies a cup can contain?

PERENNIAL MATH

TEST # 1

November 2017 Intermediate Level Answers

<table>
<tr><td>1) The third number is made by adding the previous 2 numbers. What number will go in the last box?
<table><tr><td>0.8</td><td></td><td>2.3</td><td></td><td></td></tr></table></td><td>6.1</td></tr>
<tr><td>2) The science test has some 3-point and some 4-point questions and 100 is the highest score you can make. If there are four more 4-point questions than 3-point questions, how many 3-point questions are on the test?</td><td>12</td></tr>
<tr><td>3) Mike and Bill are cousins whose ages are the same two different digits, only reversed. The sum of their ages is 132. Mike is less than 20 years older than Bill. How old is Bill?</td><td>57</td></tr>
<tr><td>4) Use the numbers 1-6 in the circles so that the sum along each side of the triangle equals 11. What is the sum of the three corner circles?
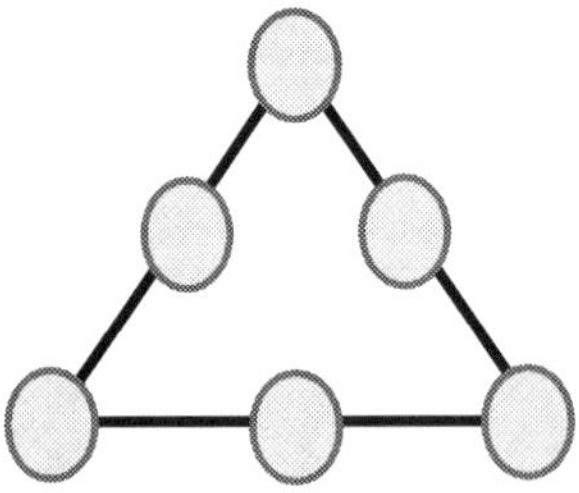</td><td>12</td></tr>
<tr><td>5) How many ways can 20 be written as the sum of 3 different numbers?</td><td>24</td></tr>
<tr><td>6) Tracy has 50 pennies and 6 cups. She plans to put an odd number of pennies in each cup. No cup can contain a number that is divisible by 5 pennies because she doesn't like that number. What is the most amount of pennies a cup can contain?</td><td>43</td></tr>
</table>

PERENNIAL MATH TEST # 1

November 2017 Intermediate Level Solutions

1) The third number is made by adding the previous 2 numbers. What number will go in the last box? First solve for the missing number in the second box: 2.3 - 0.8 = 1.5. Now you can solve for the 4th and 5th numbers with addition. 1.5 + 2. 3 = 3.8. And finally 3.8 + 2.3 = 6.1

0.8	1.5	**2.3**	3.8	6.1

6.1

2) Make a table to see how many of each type question was on the test until you find the amount where there are 4 more 4-point questions. There were twelve 3 point questions on the test.

3 point questions	0	8	12
4 point questions	25	19	16
Points x 3	0	24	36
Points x 4	100	76	64
Total points	100	100	100

12

3) Mike and Bill are cousins whose ages are the same two digits, only reversed and the sum of their ages is 132. The two digit numbers who sum to 132 are:
39 + 93, 48 + 84, and 57 + 75. Since Mike is less than 20 years older than Bill they must be 57 and 75. How old is Bill? Bill is 57.

.

66

PERENNIAL MATH TEST # 1

November 2017 Intermediate Level Solutions

4) Use trial and error to place the numbers 1-6 in the circles so that the sum along each side of the triangle equals 11. The sum of the three corner circles is 6 + 2 + 4 = 12.

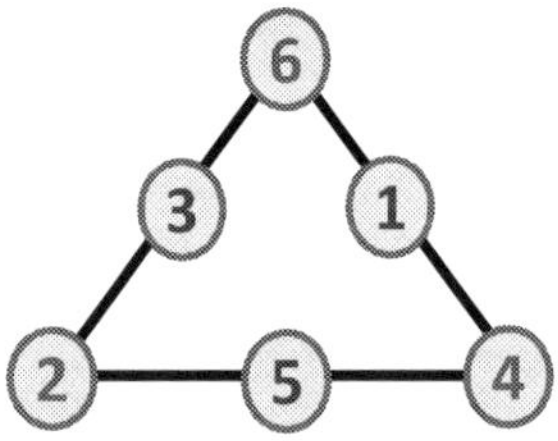

12

5) How many ways can 20 be written as the sum of 3 different numbers? Make an organized list:

1	2	17
1	3	16
1	4	15
1	5	14
1	6	13
1	7	12
1	8	11
1	9	10
2	3	15
2	4	14
2	5	13
2	6	12
2	7	11
2	8	10

3	4	13
3	5	12
3	6	11
3	7	10
3	8	9
4	5	11
4	6	10
4	7	9
5	6	9
5	7	8

24

6) Act this out. Place the lowest amount of pennies in each cup such that all are odd numbers and leave out numbers divisible by 5. The lowest odd number is 1 and the instructions did not specify that the amounts had to be different, so all the cups could contain 1 penny except that would make the last cup contain 45. That won't work because 45 is divisible by 5. So one of the cups must contain 3. Therefore, the most amount of pennies that can be in a cup is 43.

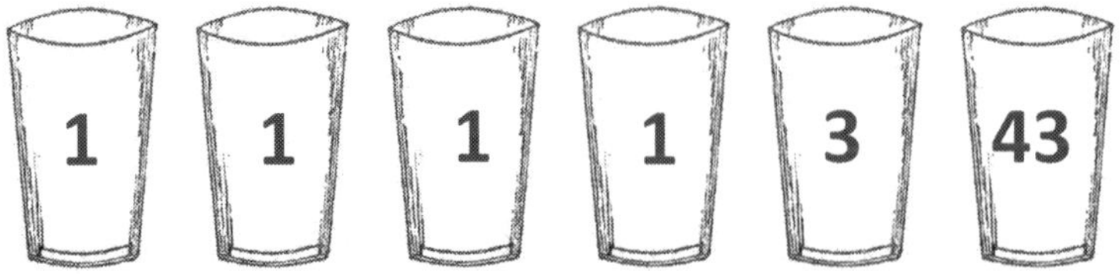

43

Name________________________________

PERENNIAL MATH

TEST #2

December 2017 Intermediate Level

1) What is the value of (156 + 263 + 539) – (56 + 163 + 239)?

2) The ratio of children to adults at the zoo is two to one. If there were 285 people in the zoo, how many were children?

3) How many squares of any size are there in this figure?

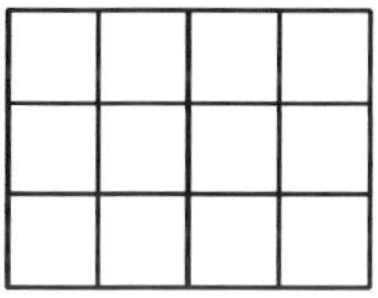

4) Two folders and one ruler cost $3.17 and two rulers and one folder costs $3.97. How much does one folder cost? Write your answer as a decimal.

5) If you rolled two dice and multiplied the two numbers, what would be the probability of rolling a product that is greater than 15? Write your answer as a fraction.

6) The sum of three consecutive numbers is 54. What is the product of those three numbers?

PERENNIAL MATH

TEST #2

December 2017 Intermediate Level Answers

1) What is the value of (156 + 263 + 539) – (56 + 163 + 239)?	**500**
2) The ratio of children to adults at the zoo is two to one. If there were 285 people in the zoo, how many were children?	**190**
3) How many squares of any size are there in this figure? 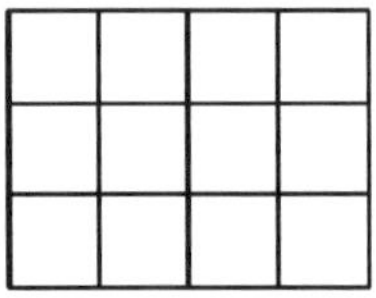	**20**
4) Two folders and one ruler cost $3.17 and two rulers and one folder costs $3.97. How much does one folder cost? Write your answer as a decimal.	**.79**
5) If you rolled two dice and multiplied the two numbers, what would be the probability of rolling a product that is greater than 15? Write your answer as a fraction.	**11/36**
6) The sum of three consecutive numbers is 54. What is the product of those three numbers?	**5814**

PERENNIAL MATH

TEST #2

December 2017 Intermediate Level Solutions

1) What is the value of (156 + 263 + 539) – (56 + 163 + 239)? Use smart subtraction to group like terms: (156 – 56) + (263 – 163) + (539 – 239) = 100 + 100 + 300 = 500

500

2) The ratio of children to adults at the zoo is two to one. There were 285 people in the zoo. To find out how many were children use a ratio of 2:1 and solve for each portion. There are 3 portions so 285/3 = 95. Twice that is 2 x 95 = 190. There were 190 children at the zoo.

190

3) Make a chart or organized list to count the number of each type of square

How many squares make the shape	Number of that type
1	12
4	6
9	2
Total	20

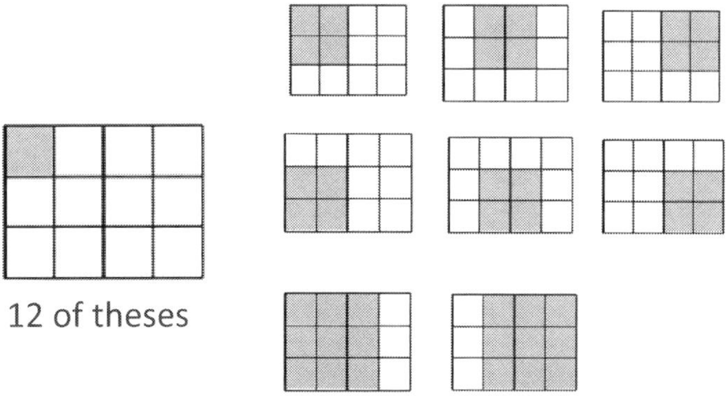

20

PERENNIAL MATH

TEST #2

December 2017 Intermediate Level Solutions

4) Use algebra. Two folders and one ruler cost \$3.17 and two rulers and one folder costs \$3.97. One folder costs .19.

$$\begin{array}{r} 2F + R = 3.17 \\ \underline{+\ F + 2R = 3.97} \\ 3F + 3R = 7.14 \end{array}$$

(Divide both sides by 3)

$$\frac{3F + 3R}{3} = \frac{7.14}{3}$$

$F + R = 2.38$ (Now subtract this from one of the original equations to get F)

$$\begin{array}{r} 2F + R = 3.17 \\ \underline{-\ F + R = 2.38} \\ F \quad = .79 \end{array}$$

.79

5) Draw a diagram to see the possibilities of products greater than 15 when the two dice are multiplied. There are 11 products that are greater than 15. The probability of rolling a product higher than 15 is 11/36.

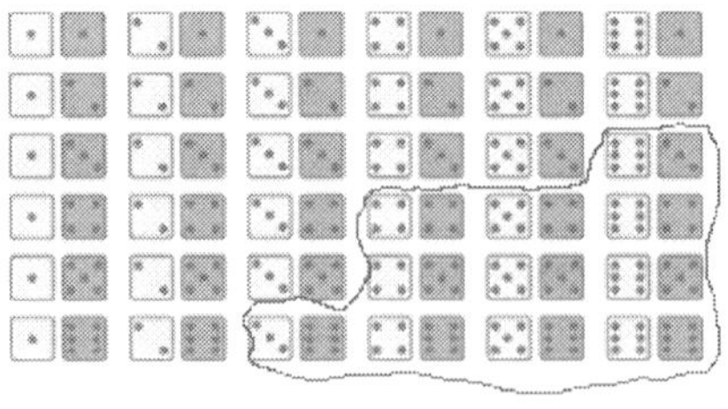

11/36

6) To find three consecutive numbers that have a sum of 54, divide 54 by 3 to get the average or middle number. 54/3 = 18. The three numbers are 17, 18, and 19. The product of those three numbers is 17 x 18 x 19 = 5814.

5814

Name________________________________

PERENNIAL MATH

TEST #3

January 2018 Intermediate Level

1) What is the sum of the largest and smallest prime number less than 100?	
2) The area of a rectangle is 48 square feet. Its width and length are whole numbers. What is the difference between the greatest possible perimeter and the least possible perimeter of this rectangle?	
3) If I add this number to 349, I get the same answer as I would if I subtracted it from 485. What is the number?	
4) The Jones Family has a swimming pool that measures 20 x 12 feet. There is a three foot wide tile walkway around the entire pool. What is the area of this tile walkway? 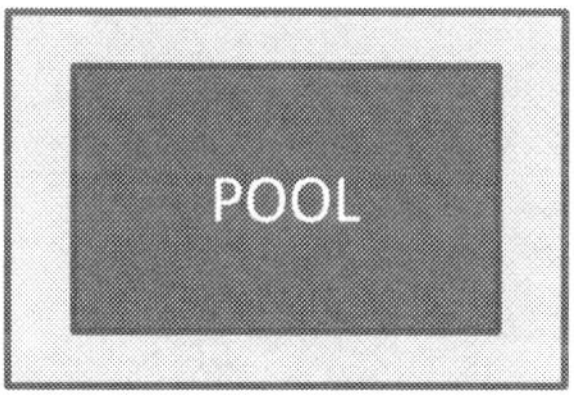	
5) John took his pennies out of his piggy bank and discovered that he could put his pennies in equal piles of three, four, five, six, seven or eight. What is the least number of pennies that he could have had?	
6) Each letter in the words PERENNIAL MATH is written on a separate piece of paper and put in a box. What is the difference between the probability that the first piece of paper drawn from the box will be a consonant and the probability that the first letter drawn will be a vowel? Write your answer as a fraction.	

PERENNIAL MATH

TEST # 3

January 2018 Intermediate Level Answers

1) What is the sum of the largest and smallest prime number less than 100?	99
2) The area of a rectangle is 48 square feet. Its width and length are whole numbers. What is the difference between the greatest possible perimeter and the least possible perimeter of this rectangle?	70
3) If I add this number to 349, I get the same answer as I would if I subtracted it from 485. What is the number?	68
4) The Jones Family has a swimming pool that measures 20 x 12 feet. There is a three foot wide tile walkway around the entire pool. What is the area of this tile walkway? 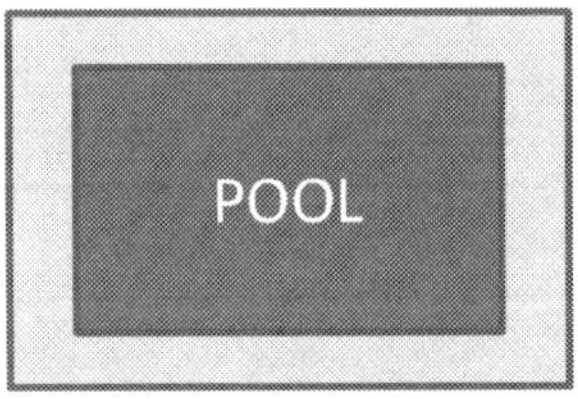	228
5) John took his pennies out of his piggy bank and discovered that he could put his pennies in equal piles of three, four, five, six, seven or eight. What is the least number of pennies that he could have had?	840
6) Each letter in the words PERENNIAL MATH is written on a separate piece of paper and put in a box. What is the difference between the probability that the first piece of paper drawn from the box will be a consonant and the probability that the first letter drawn will be a vowel? Write your answer as a fraction.	3/13

PERENNIAL MATH

TEST # 3

January 2018 Intermediate Level Solutions

1) What is the sum of the largest and smallest prime number less than 100 is
97 + 2 = 99.

99

2) Make a chart to find the greatest and least possible perimeters of a rectangle whose area is 48 square feet. Then subtract to find the difference between the greatest possible perimeter and the least possible perimeter of this rectangle.
98 ft. – 28 ft. = 70 ft.

Length	Width	Perimeter
48	1	98
24	2	52
16	3	38
12	4	32
8	6	28

70

3) If I add this number to 349, I get the same answer as I would if I subtracted it from 485. Use algebra to find the number.

$$349 + x = 485 - x$$
$$2x = 485 - 349$$
$$2x = 136$$
$$x = 68$$

Check: 349 + 68 = 417   485 – 68 = 417

68

PERENNIAL MATH

TEST # 3

January 2018 Intermediate Level Solutions

4) The entire area is 26 x 18 = 468 ft^2 and the area of the pool is 20 x 12 = 240 ft^2. Subtract the area of the pool from the entire area to find the area of the sidewalk. 468ft^2 - 240 ft^2 = 228 ft^2. Therefore, the area of the sidewalk is 228 ft^2.

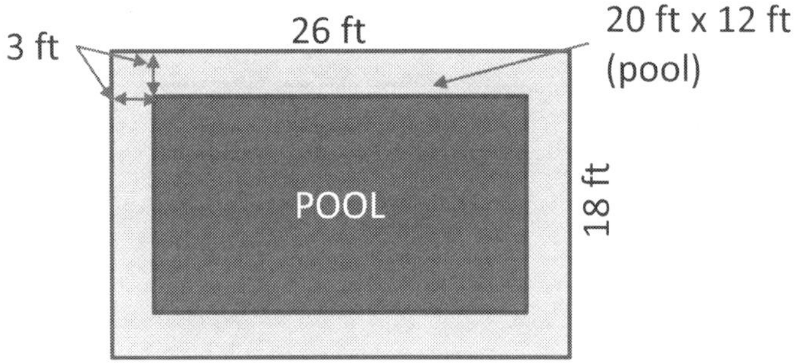

228

5) To find the least number of pennies that John could have that can be put equal piles of three, four, five, six, seven or eight you can do a prime factorization. In this list of prime factors the first number that is divisible by 3, 4, 5, 6, 7 or 8 is 840.

3 = 3
4 = 2 x 2
5 = 5
6 = 2 x 3
7 = 7
8 = 2 x 2 x 2
2 x 2 x 2 x 3 x 5 x 7 = 840

840

6) To find the difference between the probability that the first piece of paper drawn from the box will be a consonant and the probability that the first letter drawn will be a vowel in the words PERENNIAL MATH first find out the probability of consonants and then the probability of vowels. There are 8 consonants and 5 vowels in PERENNIAL MATH. Therefore the difference of the two is 8/13 – 5/13 = 3/13

3/13

Name________________________________

PERENNIAL MATH

TEST #4

February 2018 Intermediate Level

1) If $4n = 52$, find $3n + 12$	
2) The closet floor is a rectangular space completely covered with square tiles. The space is 9 tiles long and 7 tiles wide. Find the number of tiles that touch the walls of the space.	
3) Eric earned $27,000 his first year. In his seventh year he earned $34,764. If Eric's salary increases by the same amount each year, how much will he make in his twelfth year?	
4) This figure was made of cubes that are stacked on top of each other and the figure was placed on a table. If a bucket of red paint were poured over the figure, how many cubes would only have 4 red sides? 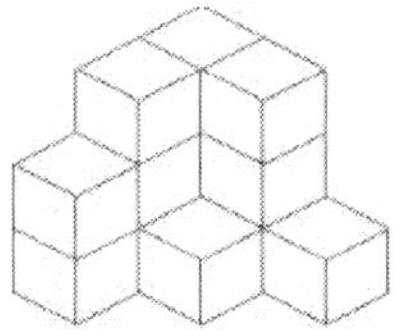	
5) Twenty eight people were surveyed about their favorite video game and 18 liked Video Game X and 16 liked Video Game Y. Four people said they didn't like either game. How many people liked both games?	
6) A bicycle rider rides at a constant speed. By 9:00 AM, he has completed 3/8 of his ride; by 10:00 AM, he has completed 3/4 of his ride. How many minutes did the whole ride take?	

PERENNIAL MATH

TEST # 4

February 2018 Intermediate Level Answers

1) If $4n = 52$, find $3n + 12$	**51**
2) The closet floor is a rectangular space completely covered with square tiles. The space is 9 tiles long and 7 tiles wide. Find the number of tiles that touch the walls of the space.	**28**
3) Eric earned $27,000 his first year. In his seventh year he earned $34,764. If Eric's salary increases by the same amount each year, how much will he make in his twelfth year?	**41,234**
4) This figure was made of cubes that are stacked on top of each other and the figure was placed on a table. If a bucket of red paint were poured over the figure, how many cubes would only have 4 red sides? 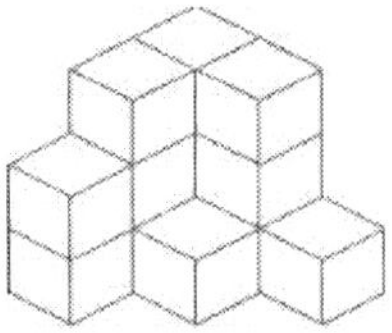	**4**
5) Twenty eight people were surveyed about their favorite video game and 18 liked Video Game X and 16 liked Video Game Y. Four people said they didn't like either game. How many people liked both games?	**10**
6) A bicycle rider rides at a constant speed. By 9:00 AM, he has completed 3/8 of his ride; by 10:00 AM, he has completed 3/4 of his ride. How many minutes did the whole ride take?	**160**

PERENNIAL MATH

TEST #4

February 2018 Intermediate Level Solutions

1) If $4n = 52$, find $3n + 12$. First solve for n. 4n = 52, so 52/4 = n. n = 13. Therefore, 3n + 12 = 3(13) + 12. 39 + 12 = 51.

51

2) Draw a rectangle that is 9 tiles long and 7 tiles wide. You will notice that each corner has a shared tile. Therefore, the number of tiles that go around the rectangle are 2(9) + 2(5) = 18 + 10 = 28 tiles that go around the rectangle.

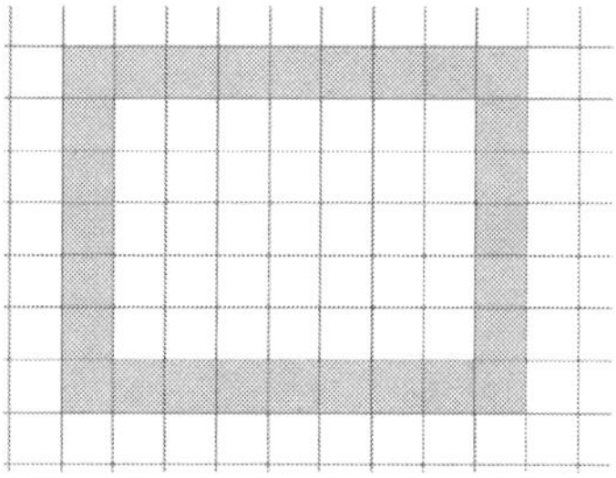

28

3) Since Eric earned $34,764 his seventh year and he earned $27,000 his first year you can subtract to find the amount of increase.

$34,764 - $27,000 = $7,764.

Divide $7,764 by 6 to find the amount of increase each year.

$7,764/6 = $1,294.

Multiply $1294 by 5 to see how much increase he will get in the next 5 years.

$1,294 x 5 = $6,470.

Add this amount to his 7 year pay

$34,764 + $6,479 = $41, 234.

If Eric's salary increases by the same amount each year, he will make $41,234 in his twelfth year.

41,234

PERENNIAL MATH — TEST #4

February 2018 Intermediate Level Solutions

4) If a bucket of red paint were poured over the figure, how many cubes would only have 4 red sides? Locate the blocks that only have another block touching two of their sides, or blocks that have one block touching their side and the table touching the bottom side. Out of the 13 blocks that make this figure, only 4 would have red paint on four of their sides.

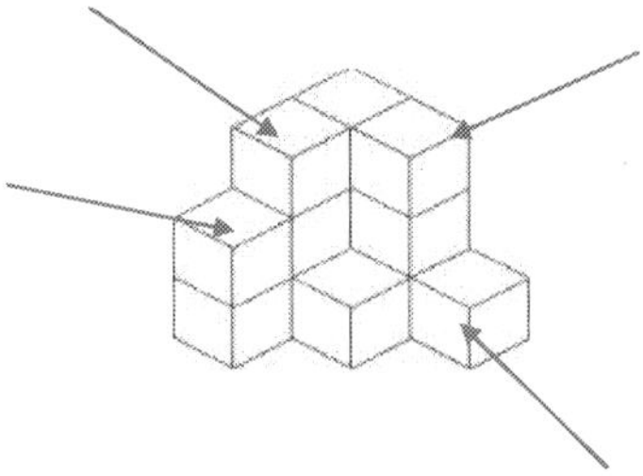

4

5) Use a Venn Diagram. Out of the 28 people, 18 like Game X, 16 liked Game Y and 4 liked neither. To find how many people liked both games solve for the regions and subtract from the total (28). Ten people like both Game X and Game Y.

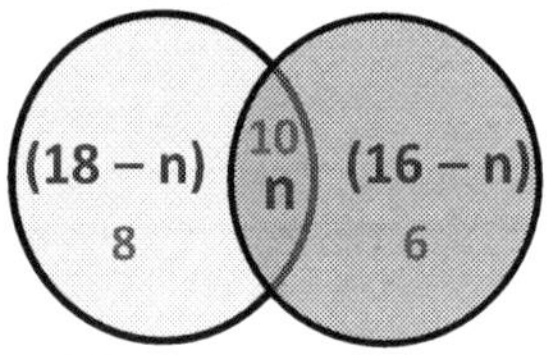

$28 = (18 - n) + n + (16 - n) + 4$

$28 = 18 + 16 + 4 - n$

$n = 38 - 28$

$n = 10$

10

6) Use algebra to solve for how many minutes his whole ride would take. By 9:00 he has completed 3/8 of his ride; by 10:00, he has completed 3/4 of his ride. Calculate how much 1 hour is equal to because you know the distance traveled from the beginning and to 10:00. He traveled 3/8 and then in one hour he was at 3/4 of the distance so $3/8 + x = 3/4$ or $6/16 + x = 12/16$. So 1 hour is equal to 6/16 or 3/8. He had to have started at 8:00 since he had traveled 3/8 of the distance by 9:00. So if 3/8 equal 1 hour, then we have to solve for what 1/4 of the distance will be since he has traveled 3/4 by 10:00. We know he has traveled 120 minutes or 2 hours. So let's solve for the entire time.

$3/4 = 120/x$

$3x = 480$

$x = 160$

160

Name________________________________

PERENNIAL MATH

TEST #1

January 2018 Intermediate Level

1) What is the sum of the two missing numbers? ______, 36, 49, 64, ______	
2) Tani brought 17 trays of cookies with 14 cookies on each tray. Jill brought 13 trays of 15 cookies each. If all of the cookies are sold for 24 cents, how much more money did Tani make than Jill?	
3) Three darts are thrown at the target with areas for 2, 3, and 4 points as shown. If all the darts hit the target, how many different point totals are possible? 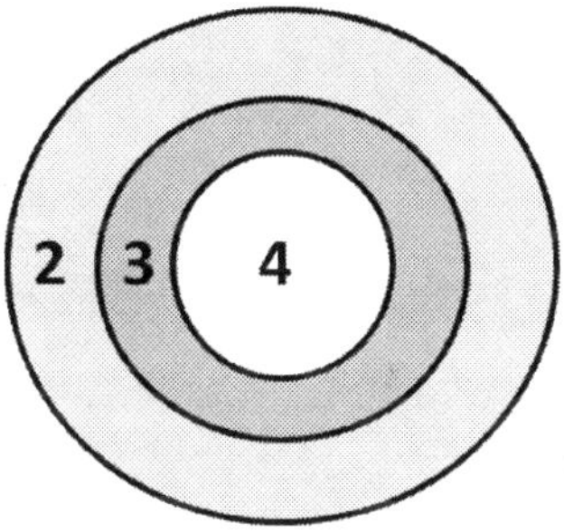	
4) There were cats and birds in the back of the pet shop. Billy counted 40 heads and 106 feet. How many birds did he count?	
5) Barry rode his bike a total of 65 miles for 5 days. Every day he rode 4 miles more than the day before. How many miles did Barry ride the last day?	
6) Each of the 39 students in Mrs. Bridge's class takes art or music or both an art and a music. Twenty students take art and 26 students take music. How many students have both art and music?	

PERENNIAL MATH

TEST # 1

January 2018 Intermediate Level Answers

1) What is the sum of the two missing numbers? _____, 36, 49, 64, ______	106
2) Tani brought 17 trays of cookies with 14 cookies on each tray. Jill brought 13 trays of 15 cookies each. If all of the cookies are sold for 24 cents, how much more money did Tani make than Jill?	10.32
3) Three darts are thrown at the target with areas for 2, 3, and 4 points as shown. If all the darts hit the target, how many different point totals are possible? 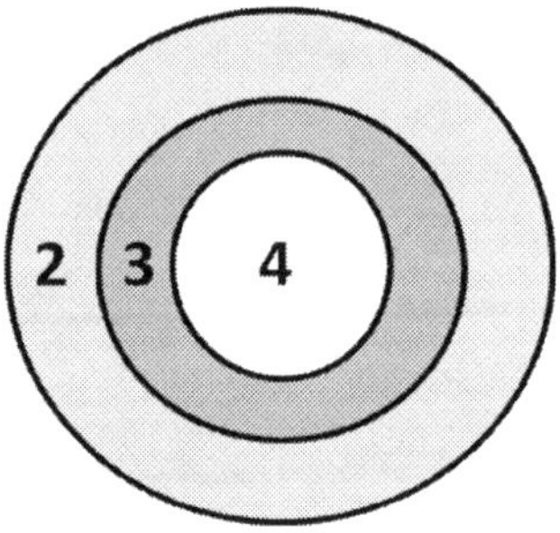	7
4) There were cats and birds in the back of the pet shop. Billy counted 40 heads and 106 feet. How many birds did he count?	27
5) Barry rode his bike a total of 65 miles for 5 days. Every day he rode 4 miles more than the day before. How many miles did Barry ride the last day?	21
6) Each of the 39 students in Mrs. Bridge's class takes art or music or both an art and a music. Twenty students take art and 26 students take music. How many students have both art and music?	7

PERENNIAL MATH

TEST # 1

January 2018 Intermediate Level Solutions

1) What is the sum of the two missing numbers? _____, 36, 49, 64, ______
The numbers in this sequence are squares. The missing numbers are 25 and 81.
Their sum is 25 + 81 = 106.

106

2) Tani brought 17 trays of cookies with 14 cookies on each tray. Jill brought 13 trays of 15 cookies each. If all of the cookies are sold for 24 cents, how much more money did Tani make than Jill?

Tani had 17 x 14 = 238 cookies. She made 238 x .24 = $57.12
Jill had 13 x 15 = 195 cookies. She made 195 x .24 = $46.80
$57.12 – $46.80 = $10.32

10.32

3) Three darts are thrown at the target with areas for 2, 3, and 4 points as shown. Make an organized list to find how many different point totals are possible. There are 7 different point totals that can be scored on the target.

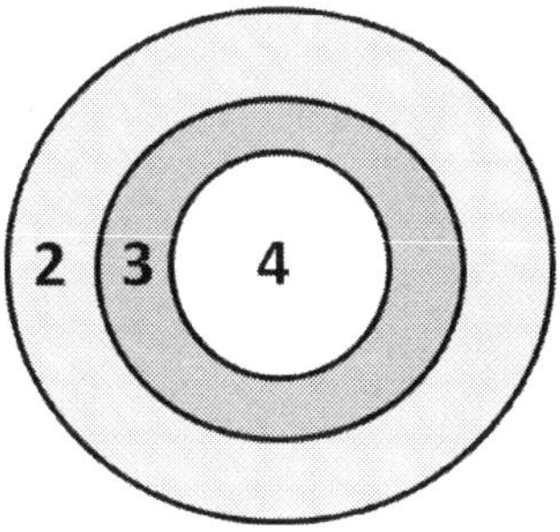

Dart 1	Dart 2	Dart 3	Points	
2	2	2	6	✓
2	2	3	7	✓
2	2	4	8	✓
2	3	3	8	
2	3	4	9	✓
2	4	4	10	✓
3	3	3	9	
3	3	4	10	
3	4	4	11	✓
4	4	4	12	✓

7

PERENNIAL MATH

TEST # 1

January 2018 Intermediate Level Solutions

4) There were cats and birds in the back of the pet shop. Billy counted 40 heads and 106 feet. To find How many birds he counted use a table. Make a chart and if you start with 20 of each animal you will see that there needs to be fewer cats. So, decrease the number of cats until you have 106 total feet. There were 27 birds.

Cats	20	19	18	17	16	15	14	13
Birds	20	21	22	23	24	25	26	27
Total heads	40	40	40	40	40	40	40	40
Cats x 4 feet	80	76	72	68	64	60	56	52
Birds x 2 feet	40	42	44	46	48	50	52	54
Total feet	120	118	116	114	112	110	108	106

27

5) Barry rode his bike a total of 65 miles for 5 days. Every day he rode 4 miles more than the day before. Use algebra to find how many miles Barry rode the last day. Let x represent the miles he rode on the first day.

$$x + (x + 4) + (x + 8) + (x + 12) + (x + 16) = 65$$
$$5x + 40 = 65$$
$$5x = 25$$
$$x = 5$$ He rode 5 miles on the first day.

So the last day is $x + 16$ or $5 + 16 = 21$ miles.
Check this by adding up the days $5 + 9 + 13 + 17 + 21 = 65$.

21

6) Each of the 39 students in Mrs. Bridge's class takes art or music or both an art and a music. Twenty students take art and 26 students take music. Use a Venn diagram and let x equal the number of students that have both art and music.

$$39 = (20 - x) + x + (26 - x)$$
$$39 = 46 - x$$
$$x = 7$$

There were 7 students who took both art and music.

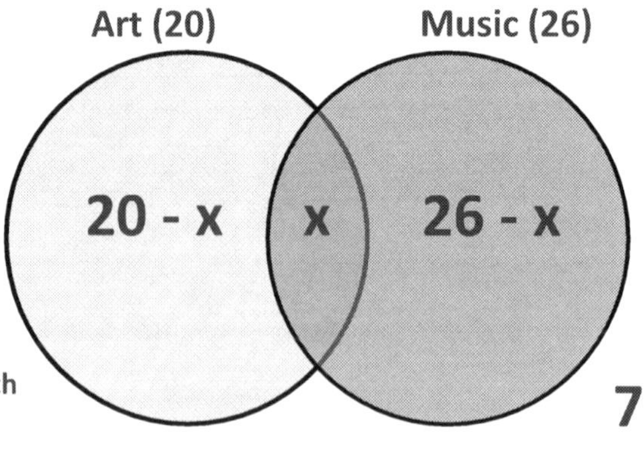

7

Name______________________________

PERENNIAL MATH

TEST #2

February 2018 Intermediate Level

1) What is the value of $200 - \frac{10}{0.1}$?

2) Six kiwis weigh the same as two oranges and two grapes. An orange weighs the same as eight grapes. How many grapes weigh the same as one kiwi?

3) Only the numbers 5 - 13 are used once to yield the column and row sums that are given in the diagram. What number goes in the center box?

6	10		24
	?		28
		13	29
27	28	26	

4) Grandpa gave 10% of 10% of his $5000 bonus check to each of his 12 grandkids. How much money did Grandpa keep for himself?

5) A photographer wants to print a photo where the length of the picture is 6 more inches than twice the height. If the total perimeter of the photo is 108 inches, how many inches are in the length of the photo?

6) Carla wrote all of the odd numbers between 300 and 700 that are divisible by both 7 and 9. What is the sum of those numbers?

PERENNIAL MATH

TEST #2

February 2018 Intermediate Level Answers

1) What is the value of $200 - \frac{10}{0.1}$?

100

2) Six kiwis weigh the same as two oranges and two grapes. An orange weighs the same as eight grapes. How many grapes weigh the same as one kiwi?

3

3) Only the numbers 5 - 13 are used once to yield the column and row sums that are given in the diagram. What number goes in the center box?

6	10		24
	?		28
		13	29
27	28	26	

11

4) Grandpa gave 10% of 10% of his $5000 bonus check to each of his 12 grandkids. How much money did Grandpa keep for himself?

4400

5) A photographer wants to print a photo where the length of the picture is 6 more inches than twice the height. If the total perimeter of the photo is 108 inches, how many inches are in the length of the photo?

38

6) Carla wrote all of the odd numbers between 300 and 700 that are divisible by both 7 and 9. What is the sum of those numbers?

2016

PERENNIAL MATH

TEST # 2

February 2018 Intermediate Level Solutions

1) What is the value of $200 - \frac{10}{0.1}$?

$200 - 100 = 100$

100

2) Six kiwis weigh the same as two oranges and two grapes. An orange weighs the same as eight grapes. How many grapes weigh the same as one kiwi?

6k = 2o + 2g and 1o = 8g.

Therefore 6k = 2(8g) + 2g (Substitute 8 grapes for 1 orange)

6k = 18g (Solve for kiwis)

1k = 3g

3

3) First complete the number in the right hand top box because 2 of the 3 numbers are given. 24 – (10 + 6) = 8.

6	10	8	24
		5	28
		13	29
27	28	26	

Now you will have 2 of the three numbers in the third column. This will allow you to solve for the missing number in that middle box. 26 – (8 + 13) = 5. The number 5 can be placed in that box.

The remaining missing numbers are 7, 9, 11, and 12. Since 5 is the only number in the middle row, the other numbers are going to need to be high in order to add up to 28. Therefore, the 11 and 12 will need to go there. Using trial and error move the remaining four numbers around until they fit to complete the sums in the rows and columns. The missing number that will be in the center box is 11.

6	10	8	24
12	11	5	28
9	7	13	29
27	28	26	

11

PERENNIAL MATH

TEST # 2

February 2018 Intermediate Level Solutions

4) Grandpa gave 10% of 10% of his $5000 bonus check to each of his 12 grandkids. How much money did Grandpa keep for himself?
10% of $5000 = $5000 x .10 = 500 and 10% of $500 is $500 x .10 = $50.
Since he gave 12 of his grandkids this amount, 12 x $50 = $600 that he gave away. He kept $5000 - $600 = $4400 for himself.

4400

5) A photographer wants to print a photo where the length of the picture is 6 more inches than twice the height. If the total perimeter of the photo is 108 inches, , how many inches are in the length of the photo? Use algebra to find how many inches are in the length of the photo. Let h = height and 2h + 6 = length. If the perimeter of a rectangle is 2(length + height) then:

2(2w + 6 + h) = 108 inches
2(3h + 6) = 108 inches
6h + 12 = 108 inches
6h = 96 inches
h = 16 inches

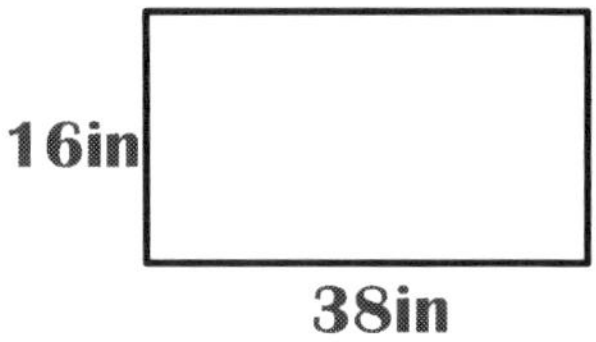

Therefore the length is 2(16inches) + 6 inches.
The photo is 38 inches long.

38

6) Carla wrote all of the odd numbers between 300 and 700 that are divisible by both 7 and 9. What is the sum of the numbers Carla wrote?
The least common multiple of 7 and 9 is 63. Therefore, numbers that are multiples of 63 will be what Carla is writing down. The first odd number greater than 300 is 5 x 63 = 315. And since she is only writing odd numbers, you can add 126 (2 x 63) to each number to get the next number.

315, 441, 567, 693.
315 + 441 + 567 + 693 = 2016 (add the numbers)

2016

Name________________________________

PERENNIAL MATH

TEST #3

March 2018 Intermediate Level

1) 17,300 – 999 = _______ + 16,000

2) The sum of two numbers is 35. The difference of these two numbers is 17. What is their product?

3) The figure below represents an 8cm by 12cm board. Each corner has a 3cm by 3cm square cut out of each corner. What is the area of the figure?

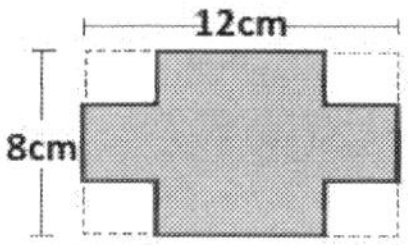

4) A truck can carry a maximum weight equal to either 18 identical large gasoline containers or 30 identical small containers. If the truck is loaded with 25 small containers, how many additional large containers can the truck carry?

5) A survey of 2000 students showed that 1400 listen to rap music, 770 listen to pop music, and 390 listen to both. What percent of the students did not listen to either?

6) Find the greatest set of three consecutive integers that are less than 30, such that the smallest is divisible by 2, the next is divisible by 3, and the largest is divisible by 4. What is the product of those three integers?

PERENNIAL MATH

TEST #3

March 2018 Intermediate Level Answers

1) 17,300 – 999 = _______ + 16,000	301
2) The sum of two numbers is 35. The difference of these two numbers is 17. What is their product?	234
3) The figure below represents an 8cm by 12cm board. Each corner has a 3cm by 3cm square cut out of each corner. What is the area of the figure? 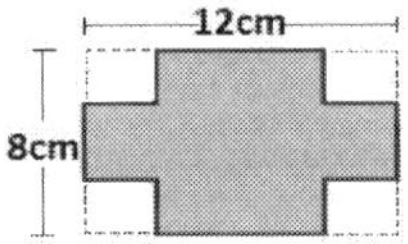	60
4) A truck can carry a maximum weight equal to either 18 identical large gasoline containers or 30 identical small containers. If the truck is loaded with 25 small containers, how many additional large containers can the truck carry?	3
5) A survey of 2000 students showed that 1400 listen to rap music, 770 listen to pop music, and 390 listen to both. What percent of the students did not listen to either?	11
6) Find the greatest set of three consecutive integers that are less than 30, such that the smallest is divisible by 2, the next is divisible by 3, and the largest is divisible by 4. What is the product of those three integers?	19,656

PERENNIAL MATH TEST #3

March 2018 Intermediate Level Solutions

1) 17,300 – 999 = _______ + 16,000
16301 = _______ + 16,000
The missing number is 301

301

2) Use algebra and let x and y represent the two numbers. The sum of two numbers is 35 can be written as x + y = 35. The difference of these two numbers is 17 can be written as x – y = 17. Find the product of the two numbers.

$$\begin{array}{r} x + y = 35 \\ + \; x - y = 17 \\ \hline 2x \quad = 52 \\ x = 26 \end{array}$$

Step 1: add equations to isolate and solve for x

$26 + y = 35$
$y = 9$

Step 2: substitute x into one of the equations and solve for y

$26 \cdot 9 = 234$

Step 3: multiply the two numbers

234

3) The figure below represents an 8cm by 12cm board. Each corner has a 3cm by 3cm square cut out of each corner. To find the area of the figure, partition the figure into smaller shapes and add up their areas to find of the area of the whole figure. So there are 2 rectangles size (12cm - 6cm) = 6cm by 3 cm. So their areas are each 18cm. 2 x 18 cm^2 = 36 cm^2. There are the other 2 rectangles size (8cm – 6cm) = 2cm by 3 cm. So their areas are each 6 cm^2. 2 x 6 cm^2 = 12 cm^2. Finally the center rectangle is 6cm x 2cm = 12 cm^2. Add all these up and the total area is 36 cm^2 + 12 cm^2 + 12 cm^2 = 60 cm^2.

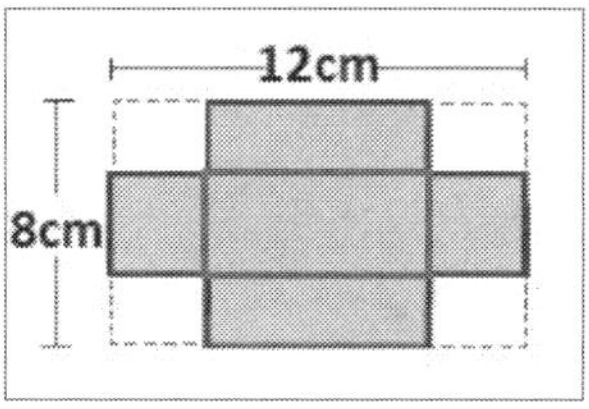

60

PERENNIAL MATH

TEST # 3

March 2018 Intermediate Level Solutions

4) A truck can carry a maximum weight equal to either 18 identical large gasoline containers or 30 identical small containers. If the truck is loaded with 25 small containers, how many additional large containers can the truck carry?
Use ratios to solve. The truck can carry a ratio of 18:30 large to small containers which is also a 3:5 ratio. So if the truck has 25 small containers, that would mean it could carry an additional 5 small containers which are equal to 3 large containers. Therefore, the truck can carry 3 additional large containers.

3

5) A survey of 2000 students showed that 1400 listen to rap music, 770 listen to pop music, and 390 listen to both. Use a Venn diagram to solve for how many students did not listen to either music and then convert to percentage.
Solve for each region:

1400 – 390 = 1010 (rap listeners only)

770 – 390 = 380 (pop listeners only)

Subtract these three regions from the total:

2000 – (1010 + 390 + 380) = 220

So, 220 Students liked neither.
Convert to a percentage

220/2000 = ____%

11% of the students liked neither.

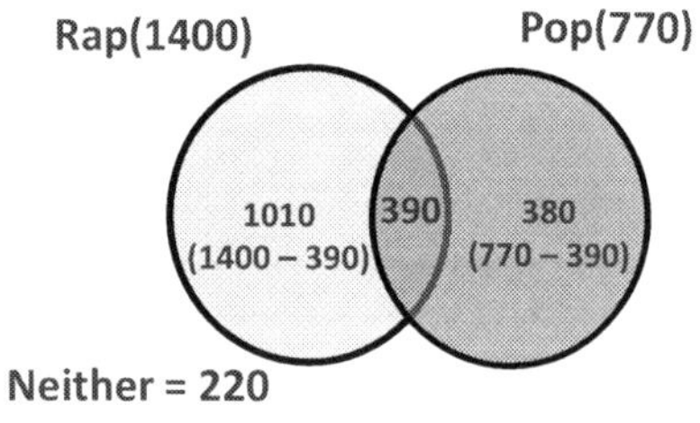

11

6) Find the greatest set of three consecutive integers that are less than 30, such that the smallest is divisible by 2, the next is divisible by 3, and the largest is divisible by 4. What is the product of those three integers?
Integers less than 30 that are divisible by 2, 3, 4 are :

2, 3, 4

14, 15, 16

26, 27, 28

Since the greatest three consecutive integers is 26, 27, and 28, find their product:

26 x 27 x 28 = 19656

19,656

Name________________________________

PERENNIAL MATH

TEST #4

April 2018 Intermediate Level

1) 5 + 4 x 8 – 2 x 2 =

2) Donald is the eleventh person counting from the beginning and the end of the line. How many people are in line?

3) Melanie divided her beads into two piles that had an equal number of beads. She gave her sister one-half of one-third of one pile. Melanie had 77 beads left. How many beads did Melanie start with?

4) The groups of pattern blocks are placed in rows with the lengths of each row given. Given the information below, how many centimeters are in the width of one diamond?

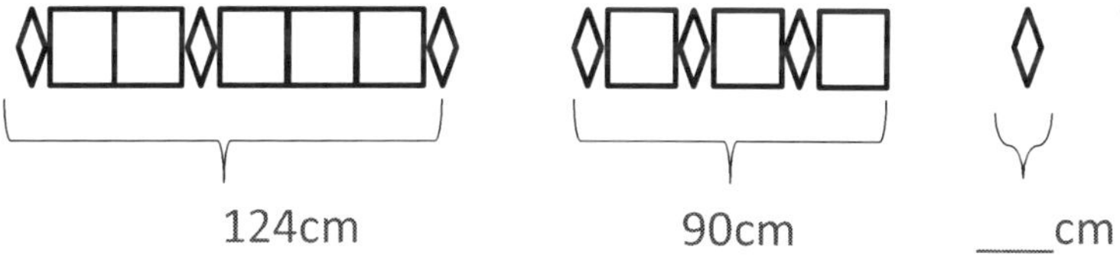

5) A cubic number is of the form n^3 for some natural number n. For example, 8 is a cubic number because $8 = 2 \times 2 \times 2 = 2^3$. How many cubic numbers exist from 1 to 2000?

6) Leon earned the following scores on seven math tests: 88, 95, 84, 100, 86, 99, 94. What does Leon have to score on the next test to have a 93 average?

PERENNIAL MATH

TEST # 4

April 2018 Intermediate Level Answers

1) 5 + 4 x 8 – 2 x 2 =	33
2) Donald is the eleventh person counting from the beginning and the end of the line. How many people are in line?	21
3) Melanie divided her beads into two piles that had an equal number of beads. She gave her sister one-half of one-third of one pile. Melanie had 77 beads left. How many beads did Melanie start with?	84
4) The groups of pattern blocks are placed in rows with the lengths of each row given. Given the information below, how many centimeters are in the width of one diamond? 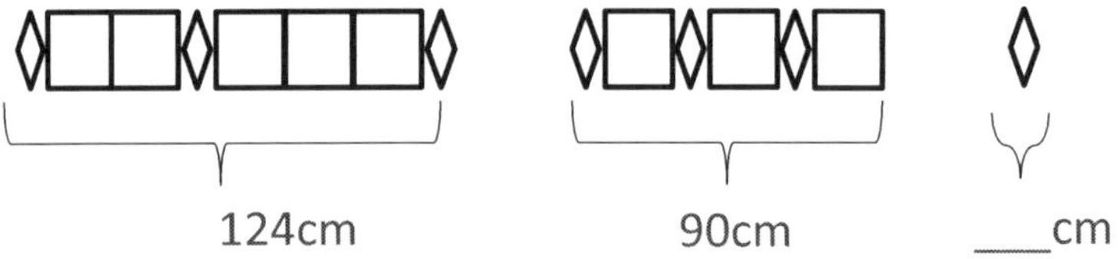	13
5) A cubic number is of the form n^3 for some natural number n. For example, 8 is a cubic number because $8 = 2 \times 2 \times 2 = 2^3$. How many cubic numbers exist from 1 to 2000?	12
6) Leon earned the following scores on seven math tests: 88, 95, 84, 100, 86, 99, 94. What does Leon have to score on the next test to have a 93 average?	98

PERENNIAL MATH

TEST #4

April 2018 Intermediate Level Solutions

1) 5 + 4 x 8 – 2 x 2 =

Use order of operations to solve:

5 + (4 x 8) – (2 x 2)

5 + 32 – 4 = 33

33

2) Donald is the eleventh person counting from the beginning and the end of the line. Draw this out to see how many people are in line. This would show that there were 10 people in front and behind Donald. So there were 21 total people in line.

21

3) Melanie divided her beads into two equal piles. She gave her sister one-half of one-third of one pile. Melanie had 77 beads left. To find how many beads Melanie started with draw a picture. Since Melanie gave away 1/6 of 1 pile, she would have given away 1/12 of all the beads since the other pile is worth 6/6. Or she would have 11/12 left 77 = 11/12 so 11/12 = 77. Each portion had to be equal to 7 beads. Therefore, Melanie had 84 beads to start with.

11x = 77

x = 7

12 ▪ 7 = 84.

Check this out. If she had 84 beads and gave 1/6 of ½ away then she gave away 1/12. She would have given away 7 beads. 84 – 7 = 77.

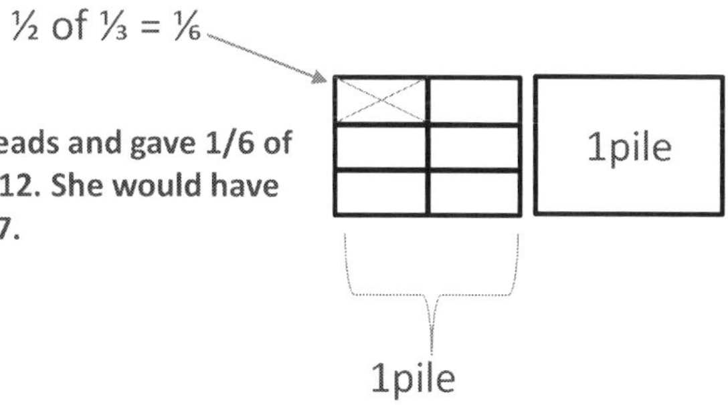

84

PERENNIAL MATH

TEST # 4

April 2018 Intermediate Level Solutions

4) To find out how many centimeters are in the width of one diamond do a comparison. If 3 squares + 3 diamonds = 90cm, then subtract this from the 5 squares + 3 diamonds = 124cm. That leaves 2 squares equal 34cm. The length of 1 square is 17cm. So substitute this back in the equation of 3 squares + 3 diamonds = 90cm. 3(17cm) + 3 diamonds = 90cm. 90cm – 51cm = 39cm. Therefore 39cm is the length of 3 diamonds, so one diamond equals 13cm.

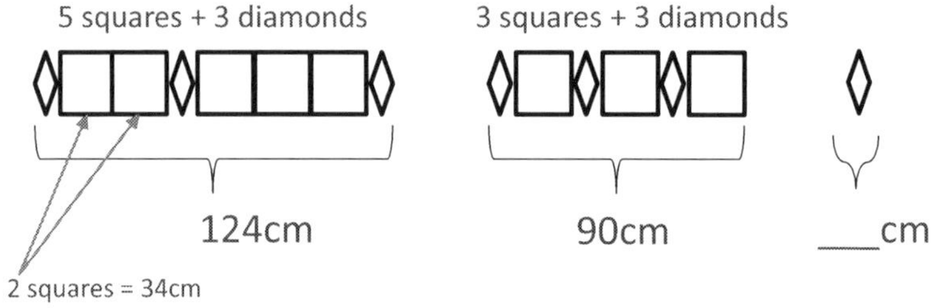

13

5) How many cubic numbers exist from 1 to 2000?

$1^3 = 1 \times 1 \times 1 = 1$

$2^3 = 2 \times 2 \times 2 = 8$

$3^3 = 3 \times 3 \times 3 = 27$

$4^3 = 4 \times 4 \times 4 = 64$

$5^3 = 5 \times 5 \times 5 = 125$

$6^3 = 6 \times 6 \times 6 = 216$

$7^3 = 7 \times 7 \times 7 = 343$

$8^3 = 8 \times 8 \times 8 = 512$

$9^3 = 9 \times 9 \times 9 = 729$

$10^3 = 10 \times 10 \times 10 = 1000$

$11^3 = 11 \times 11 \times 11 = 1331$

$12^3 = 12 \times 12 \times 12 = 1728$

There are 12 cubic numbers from 1 – 2000.

12

6) Leon earned the following scores on seven math tests: 88, 95, 84, 100, 86, 99, and 94. To find what Leon has to score on the next test to have a 93 average, first find out what his points have to add up to in order to have a 93 average on 8 tests. 93 x 8 = 744 points that must be scored to have a 93 average. Now subtract his current point total. (88 + 95 + 84 + 100 + 86 + 99 + 94 = 646). 744 - 646 = 98. Leon will need to score a 98 on the next test.

98

Made in the USA
Monee, IL
05 November 2026

67811016R00033